Edison Pascal

Biodiversity in Agriculture and Farming

Edison Pascal

Biodiversity in Agriculture and Farming

Ecological Management

ScienciaScripts

Imprint

Any brand names and product names mentioned in this book are subject to trademark, brand or patent protection and are trademarks or registered trademarks of their respective holders. The use of brand names, product names, common names, trade names, product descriptions etc. even without a particular marking in this work is in no way to be construed to mean that such names may be regarded as unrestricted in respect of trademark and brand protection legislation and could thus be used by anyone.

Cover image: www.ingimage.com

This book is a translation from the original published under ISBN 978-620-0-04876-9.

Publisher:
Sciencia Scripts
is a trademark of
Dodo Books Indian Ocean Ltd. and OmniScriptum S.R.L publishing group

120 High Road, East Finchley, London, N2 9ED, United Kingdom
Str. Armeneasca 28/1, office 1, Chisinau MD-2012, Republic of Moldova, Europe
Printed at: see last page
ISBN: 978-620-7-00280-1

"Even if we never think about it, it is good to know that our present way of living, totally oblivious to the Earth's ecological processes, will undoubtedly lead to our own extinction. We need the Earth, it does not need us. The day we cease to exist as a species, life on the planet will simply go on.

Edison Pascal

Acknowledgements

To my wife, for encouraging me and being part of this research.

To my family, children, parents and siblings.

To my students and research group.

To the N ueva Venezuela walking community, for the experience and support provided during each visit and each exploration.

To Prof. Le rm ith Torres for providing images of some species of ornithological fauna for this work.

To the University of M atanzas and the Estación Experim ental de Pastos y Forrajes "Indio Hatuey" (Cuba) for the training provided in the agroecological field.

To the University of Zulia, Faculty of Veterinary Sciences. To the Centro de Investigaciones Educativas de B iología y Q uím ica (CIEBYQ) and the Universidad Nacional Experim ental "Rafael M aría Baralt" (UNERM B) where I have a working, academic and fraternal life.

Edison Pascal

TABLE OF CONTENTS

INTRODUCTION

Biodiversity is, according to the International Convention on Biological Diversity, the term referring to the wide variety of living things on earth and the natural patterns that shape it, i.e. it is the result of millions of years of evolution according to natural processes and also of the increasing influence of human activities. Biodiversity also includes the variety of ecosystems and the genetic differences within each species that allow for the combination of multiple life forms, whose mutual interactions underpin the sustenance of life on the planet.

Environmental protection, conservation and rational use of renewable natural resources form an essential part of the state's land-use planning policy. In this sense, environmental planning and land use planning must propose strategies or methods that allow the territory to be evaluated from physical, natural, biotic, social, cultural and economic aspects, including the spatial and evolutionary character of each variable, as expressed by Berroteran, (2011).

Ecological planning (EMP) is an environmental policy instrument to regulate land use and promote sustainable development, and aims to maximise consensus between sectors and minimise environmental conflicts over land use. Ecological planning should be seen as a continuous, participatory, transparent and methodologically rigorous and systematic planning process (Sem arnat, 2009).

The ecological management process starts with the signing of a coordination agreement in which the following commitments will be established: To integrate the ecological management committee, ensuring the representation of the public, private or social sectors. G enerate the ecological management model and

the ecological strategies that will be part of the ecological management programme, establish the environmental logbook.

With ecological planning, the aim is to promote an environmental planning scheme geared towards sustainable development; within this scheme, the linking and integrality of decision-making on issues that affect the pattern of occupation of the territory will be promoted, as well as the participation of society and transparency in environmental management, as expressed by Sem Arnat, (2010).

The ecological management programme should include zoning, ecological guidelines for each zone and a set of ecological strategies. An ecological guideline is defined as a general statement reflecting the desired stage or state of a territorial unit or zone. To be valid, an ecological guideline must meet the following conditions: (1) be established by consensus among the sectors involved within the ecological management committee, (2) be based on the best technical and scientific information, and (3) be generated through systematic and transparent procedures.

For their part, the ecological strategies will include all the actions, programmes and projects of the three levels of government that seek to achieve the corresponding guidelines. Biodiversity must be taken into account when carrying out an ecological planning study, which is why it is essential to know the existing biological diversity within the territory. The conservation of biodiversity is one of the indispensable aspects to be considered in the

The ecological management of the territory due to the potential conflicts that may arise between sectors, the one representing the conservation of diversity being one of the most important, according to Rodriguez, Sem arnat (2010).

In order to properly conserve biological diversity, it is necessary to locate the richest areas, those containing the greatest diversity, those with high levels of endemicity in biological groups and those containing species that are on official lists of species to be protected.

There is an urgent need to identify critical areas for biodiversity conservation, given the high rates of deforestation and changes in land use that are currently occurring, leading to the loss of significant numbers of species as well as particular habitats and loss of functionality of biological systems. Biodiversity can be lost significantly in part due to a lack of strategies to identify what is a real priority for conservation (Rodriguez, Sem arnat, 2010).

Biological diversity (biodiversity) is the name given to all living organisms that occupy the earth and their interaction with each other. The existing biodiversity on the planet is the consequence of evolutionary processes over millions of years; each organism has its own particular form of life, which is perfectly related to its ecological niche (Pascal, 2015).

The diversity of the biological community is a function of the number of taxa and the proportional abundance of species. Diversity tends to decrease in disturbed environments as a result of the decrease in the number of taxa and the different distribution of abundance (a few taxa are very abundant). There are different expressions to measure the diversity of organism s, one of the most used is the M argalef index (M argalef, 1974).

After the deductions made above, it became evident that in the rural settlement of Nueva Venezuela in the municipality of Cabimás there is: lack of knowledge of the biodiversity that inhabits the community, lack or absence of ecological territorial planning, little knowledge of environmental and agro-ecological issues, poor conditions of access roads to the community, lack of

The lack of some basic services, such as electricity, domestic gas and drinking water, and the poor state of the dwellings in the community. According to the weaknesses detected, the hypothesis of the research arises:

Hypothesis

If the diagnosis of biodiversity in the rural settlement "Nueva Venezuela" is carried out, it will be possible to make sustainable and equitable use of the community's territory by proposing the implementation of ecological planning .

O bjectives of the research

General Objective

Identify the biodiversity associated with the Nueva Venezuela farming settlement and propose an ecological management plan for the community.

O bjectives Specific Objectives

To carry out a diagnosis to determine the biodiversity within the rural community.

Determine the importance of biodiversity and agroecology as a means of sustainable development for the inhabitants of the sector.

CHAPTER 1

Research Background

Biodiversity in Venezuela

Venezuela, thanks to its geographical location, has developed a series of landscapes and life zones where a great variety of flora and fauna are found, it has a climate that benefits its entire population, it also has natural products such as oil, as well as a very important resource such as fresh water, which in turn makes the existence of flora and fauna possible, forming a process of adaptation among the communities.

Today, this great biodiversity is threatened by human activities, such as the felling and burning of trees and by industrialisation, where man has, over time, put many species in danger of extinction, some of which have been exploited by agriculture, others by hunting and fishing.

Its importance lies in: limiting overpopulation, as not all species can survive in the same geographical space; promoting culture by identifying each state as having a different fauna and flora characteristic of each region, for example, the state has mountain ranges, represented by the frailejon and the condor.

With the accelerated and unsustainable growth of the population and the consumption of natural resources, damage has been done to our biodiversity, large industries, the transformation of life in the countryside/city, and the lack of awareness of each individual has led to the extinction of fauna and flora worldwide, producing environmental changes such as global warming.

It is worth noting that humans derive benefits from biodiversity in the form of food security, health maintenance, energy security and water,

m aterials, recreation, spiritual satisfaction, artistic inspiration, among others. Therefore, the existence of biodiversity guarantees a good quality of life for human beings and is a decisive factor for their very survival.

Changes in the interactions between species can also have negative effects on ecosystem processes. Thus, the loss of an essential species can alter the services that these systems provide to humans. Biodiversity itself provides food, fresh water, fertile soils for survival, medicines and textile fibres to care for us, to clothe us, other raw materials. Healthy ecosystems regulate our climate and absorb CO_2 , purify the water we drink, control flooding and slow erosion, protect and fertilise the soil that sustains our food.

Landscapes or natural areas are an important part of our natural heritage and culture, offering, among other values, a healthy environment for leisure, recreation, peace, tranquillity, discovery and learning. In short, the quality of life depends on the maintenance of habitats, living and healthy ecosystems. The loss of biodiversity leads to a reduction in the benefits that humans obtain from it, and also increases the risk, at a speed never seen before, of surprises such as the extinction of species, changes in climate, among others. G il, et. al., (2003).

1.2. Identification of Priority Conservation Sites

According to the International Union for Conservation of Nature (IUCN), the global record of extinct species is at least 785, with some 65 species surviving only in captivity or in the domestic state, a figure fifty times higher than the natural rate of extinction (IUCN 2009). In the most recent assessment, nearly 200 species were added to the list of 16,306 species threatened with extinction. In terms of

Overall, one amphibian in four, one bird in eight, one third of amphibians and 70 % of plants are endangered (IUCN 2009). This is one of the greatest environmental alerts: we are facing the possibility of a new phenomenon of mass extinction, this time generated by a single species, *Homo sapiens*.

In this context, Venezuela, the thirty-third largest country in the world in terms of surface area, has nearly 16 000 plant species, 351 mammal species, 1 361 bird species, 341 reptile species, 284 amphibian species, 1 000 freshwater fish species, 791 marine and brackish water species, which has given it a privileged position among the countries with the greatest number of species in the world (Aguilera et al. 2003, Rodríguez and Rojas-Suárez). This has given it a privileged position among the most biodiverse countries with the highest number of species in the world (Aguilera et al. 2003, Rodríguez and Rojas-Suárez, 2008). In contrast, of the 5 067 threatened species present in South America, 234 are found in Venezuela.

These, in relation to the 2,052 at risk, place the country in fifth place among the 14 American countries with the most species at risk (IUCN, 2009). Information on the extinction of species in Venezuela has increased notably with the publication of the red books of fauna and flora (Llam ozas et al. 2003, Rodríguez and Rojas-Suárez, 2008). However, elucidating the origins of this problem is difficult, as information is scattered and imprecise.

It was only in the 19th century that some cases began to be documented in detail, and it is only since then that an approach to the main causes of extinction and a more precise knowledge of the species studied has been possible. The first evidence of extinction processes is the destruction of the Cubagua pearl bank at the beginning of the 16th century. During Columbus' third voyage in 1498, the wealth of pearls produced by the Pinctada im bricata oyster was reported from the coasts of the island of Cubagua, which quickly led to its colonisation.

It has been estimated that between the years 1515 and 1542 at least 11 326 kilos of pearls were harvested, approximately 113 260 000 oysters, which when added to those not reported and after adjusting calculations, would have represented one billion oysters harvested in 30 years. This high harvest rate depleted the resource. Today, almost five hundred years later, the pearl beds have still not recovered, and there is speculation that the *Arca zebra* pepitona may have com petitively displaced the pearl oyster (SARPA, 1996).

By the 17th and 18th centuries the use of fauna and flora as a resource was little documented, although it is estimated that indiscriminate exploitation intensified. An example of this situation can be deduced from the pattern of extinction of psittacines in the insular Caribbean, where it is estimated that at least 16 species of macaws, parrots and parakeets became extinct in this period, equivalent to 50 % of the region's psittacines (Snyder et al. 1987).

1.3.   Im plem ented Solutions

One of the main difficulties for the conservation of endangered species was the lack of precise references that would make it possible to discriminate which species really required conservation action and which did not require urgent measures. At the beginning of the 20th century, all game fauna was often reported as being at risk, but some of the species were not necessarily considered as endangered, such as mule deer, macaws and pumas, classified on a par with other animals that do classify as endangered species, including caimans.

It should be noted that the red lists are a good starting point, but more direct action is required to reverse a threatening situation, and in this respect Venezuela has promoted several legal instruments, created a strong

The project also promoted the implementation of projects for the management and conservation of endangered fauna with the aim of recovering their populations.

In Venezuela, these general actions range from the creation of the first Ministry of the Environment in Latin America in the 1970s, to the conception of specialised institutions that set the standard in wildlife matters, such as the now defunct Autonomous Profauna Service. In addition to a comprehensive set of general environmental laws, there has been legislation for endangered species.

1.4. Current Situation of the Ecological O rdenam ent and Identification of the Biodiversity of the Farm Settlement

Traditional agricultural activities have been developed with their back to the conservation of the environment, biodiversity and social differences, it is necessary in the productive rural sectors to carry out an ecological territorial planning that benefits the community, the existing biodiversity in the sector, carrying out sustainable and environmentally friendly agricultural activities. The rural settlement of Nueva Venezuela has sufficient land and water resources to be able to develop agricultural and livestock plans that are environmentally friendly, but there is a lack of knowledge about biodiversity and ecological planning.

G arzas Reales (*A rdea alba*) near the lagoon of the Asentam iento Cam pesin o "N ueva V ene zuela" peri-urban area of the city of Cabim as. Source: The author.

This research was aimed at highlighting the importance of recognising natural spaces in the current global context of socio-spatial and territorial changes, based on the experience of observation, the aim is to analyse the state of biodiversity in the "Nueva Venezuela" Rural Settlement and to determine how it can be used.

In addition, to be able to detect the benefits that the application of ecological agriculture would have for the community, as it will allow them to develop a tool for the supply of food produced by themselves, and also to inform the inhabitants of the great importance of the existing body of water in the locality, as it can generate benefits at an economic level in terms of the use of aquaculture.

The analysis of a territorial space is complex, given that there are different approaches to its study, for the development of this research special attention has been paid to the settlement of populations in forested areas and other areas of globalisation, and it cannot be otherwise, without analysing the territory under study as an isolated space; rather

The different scales and areas that make up and influence it should be analysed from the different scales and spheres. There are areas of the world that are completely abandoned to their fate, the demographic population increase, social, environmental or economic problems have led to the population of environmental areas where an impact is exerted on the characterisation of the environment and its natural components.

Territory is an important element in identity, because it locates populations, offers possibilities, establishes certain conditions. The interaction between society/territory over time gives rise to areas with very unique ways of life, customs and organisation of space; the city of Cabimbas is a sample of the different settlements that man has made throughout its extension in the different sectors, neighbourhoods, communities and urbanisations.

Ignorance of the biodiversity existing in a territory for agricultural use could trigger its loss and the benefits that bio logical diversity could have in an integrated agricultural system; however, the lack of ecological management places this biodiversity at risk, this being a fundamental aspect to avoid potential conflicts that could arise in the community. The immediate need to make rational use of environmental resources and the urgent need to halt the deterioration and degradation of ecosystems in Latin America, specifically in Venezuela, makes the study of ecosystems a priority, with the aim of proposing better spatial planning within a socio-economic and ecological environment (Berroteran, 2001).

According to the deductions made above, it can be deduced that the following situations were specifically detected: lack of knowledge of the biodiversity that inhabits the unit, lack or absence of ecological territorial planning, little knowledge of environmental and agro-ecological issues, poor condition of the access roads to the unit, lack of some basic services, such as electricity, domestic gas and drinking water, poor condition of the houses in the unit, these were the main reasons for the lack of basic services, such as electricity, domestic gas and drinking water; the poor condition of the houses that are located in the unit.

The selected community has the highest priority problems that need to be solved. Through the implementation of ecological management and the identification of biodiversity in the rural settlement of Nueva Venezuela, it will be possible to make sustainable and equitable use of the community's territory.

1.5.   Intuitive and Quantitative Methods for Identifying Priority Areas for Conservation

Methods for identifying priority or critical areas for conservation can have different approaches, ranging from the intuitive to the quantitative analytical. Both approaches have been used for the identification of areas containing certain attributes of conservation interest, such as the presence of flagship or endangered species, or the existence of particular habitats, as relevant as an oasis within a forest.

Intuitive approaches, which are the ones that have been most commonly used in the analysis to identify critical areas for conservation, have been based on relatively simple analyses, largely based on the experience of experts. These approaches, although interesting, have a high uncertainty and a high margin of error in the delimitation of priority areas, which makes them unattractive in the context of ecological management. Given the subjective nature with which each expert can evaluate critical areas, the results can become unrepeatable and inconsistent.

On the other hand, the great development of quantitative techniques for the delim itation of critical areas for the conservation of biological diversity now makes it possible to reduce the uncertainty and inconsistency of the results. There are now different statistical approaches that allow for quantitative analyses of the biological information available for the determination of critical areas.

or critical for conservation. Likewise, the development of technologies for spatial analysis of information, together with the development of statistics for spatial analysis, has allowed the generation of predictive models of species distribution or of particular habitats at different spatial scales.

The development of Geographic Information Systems and statistical analyses through the generation of probabilistic models allow the analysis of information at regional scales, based on sampling methodologies at local scales. This approach is definitely the most convenient in the context of the ecological management of a region.

Predictive models can help prioritise surveys, identify priority or critical areas for conservation and improve conservation programmes in a region. Once the areas containing the highest specific richness of the groups of interest or where species of conservation interest coincide (such as endemic species or those categorised on the U ICN species protection list or species considered key or ecologically relevant) have been located, ecological and statistical analyses can be carried out to allow for a better design of critical areas for conservation (U ICN, 2009; PROVITA, 2009).

Some consideration needs to be given to the elements used to determine whether a site is a priority for conservation. For the identification of priority sites, it is necessary to define the key elements that should currently be used to consider them as a priority. Firstly, as representatives of biological diversity we should consider all biological groups in an area and thus have a quantitative, numerical measure that represents all of them, but in practice this is hardly possible.

Depending on the region, for some groups there is a higher level of information than for others, and for some groups the information is very poor if not non-existent (e.g. some hyperdiverse groups such as arthropods, fungi). Thus, although all biological groups make up the biological diversity of an area.

It is important to consider that species or groups of species for which an adequate level of information on their ecology and biology is available and which can be relatively easily sampled should be used for the analyses and statistical modelling. This information will allow the models to be robust, and environmental variables that appear significant in the models can be considered as ecologically relevant variables for the species.

Secondly, it can be noted that there are two approaches to identifying priority sites for conservation. One is to consider the total richness of each biological group for which information is available and the total richness of all groups. In this way we could include the sites containing the highest biological diversity.

The other approach is to consider individual species, where only species of particular conservation interest (such as species on the official IUCN protection list, endemics, relevant rare species, keystone species, i.e. those that are considered relevant for one of their attributes in the biological system and that are recognised nationally and internationally as species of conservation concern) can be included.

In any case, be it group richness, total species richness or individual approximation of relevant species, the next step is to measure the environmental characteristics of sites with higher richness or species m odels to determine the environmental variables that condition the distribution of these particular groups or species. This is achieved by generating probabilistic models that show in essence the selection of habitats that are favourable to the species for their presence. The last step is to choose the method that will allow the most optimal selection of priority sites for conservation (M assiris, 1993).

1.7. Statistical M odelation of Species Distribution and Species Richness

It is necessary to select probabilistic predictive models that can be used to determine species distribution patterns. There are a number of important statistical models for spatial analysis of information, including generalised linear models, generalised additive models, classification and regression trees, Bayesian models and canonical correspondence analysis. For this chapter we will present information on three types of modelling, with special reference to the use of generalised linear models because of their wider use, their greater statistical development and the robustness of the models obtained:

- Generalised Linear Models (GLM)
- C la s s if ic a tio n a n d regression trees
- Bayesian models

Each of the models is described below:

G ene ralised Linear Models. These are statistical models that allow the distribution of species to be predicted by obtaining a probabilistic model. These models are obtained or fitted by including environmental variables that may or may not have a normal distribution. Although they are correlative models, if the adjustment of environmental variables is carefully followed in such a way that they are adjusted variables that are ecologically significant, it is possible to have models in which the variables

represent the environmental or ecological conditions for species to occur at a site.

Thus, if these constraints do not exist in an area, the probability of the species not occurring in that area will increase. The environmental variables used by these models to predict the distribution of species or groups of species can be taken from different sources and at different scales. From variables taken directly in the field at the local scale (vegetation structure, cover, soil texture), through digitised imaged maps, to those variables that can be extracted from remote sensing (aerial photography, satellite imagery). Species distribution can be modelled in terms of species presence/absence or abundance.

Classification and regression trees. These are methods that partition the space in a m ultidim ensional way defined by the explanatory variables that are located in nodes that represent the responses of the species to the environmental variables considered, in such a way that a binary response is obtained, which is the decision represented in the tree.

Each node branch can be described as the response to the environmental variable by distribution (categorical response) or by its average value of occurrences (quantitative response) of the response variable (Toledo, 1998). Although the predictive power of these models is similar to that obtained by G LM models, they are less robust models in terms of their ecological explanation and because the terminal nodes of the tree may have few cases, which makes the models less robust (Escobar, 1997).

Bayesian models. They are based on Bayes' theorem, and combine a priori characterised probabilities of observed species or units with their probabilities of occurrence conditional on the value or class of values.

established for each environmental predictor. Conditional probabilities may be the relative frequencies of species occurrence within discrete classes of a nom inal predictor (those taken in the field). A priori probabilities may be based on previous studies reported in the literature.

Models such as the Bayesian model are included among the classification and ordination methods. The growing interest in predicting ecological processes makes Bayesian models a relatively easy method to perform and their use is increasing to achieve predictions considering the uncertainties that exist in the observations. Hierarchical formulations are very useful for prediction and can incorporate scientific judgement in a consistent probabilistic way (Delgadillo and Torres. 2008).

Bayesian statistical inference provides an alternative way of analysing data that is likely to be more appropriate for biodiversity conservation than traditional statistical methods. Bayesian methods calculate the probability of occurrence of various data values within a given parameter. Bayesian analyses are relatively easy to explain and automatically include the uncertainty of the estimate and expose a better likelihood of the representation of population status.

On the other hand, uncertainty in the model can be formally incorporated into the results of the analyses and uncertainty can be reduced by incorporating additional information in a formal and transparent way, including combining different types of data, subjectively using information from a population. Ostensibly, the fact that Bayesian methods have not been widely used is attributed to the difficulties they can present in terms of convenience. However, these problems have been rapidly disappearing due to the new solutions available in terms of numerical methods of integration and the increase of faster computers (Plonczak, 1 999).

1.8. Procedure for the Selection of Critical Areas for Conservation.

The objective of the problem in this case is to delimit or select critical areas for conservation in a given area or region through quantitative methods and with a repeatable methodology that reduces selection errors. To achieve this objective using statistical modelling as a repeatable quantitative method, a series of steps can be followed as described below:

1. The generation of statistical models requires information on environmental variables that are important in determining the distribution patterns and abundance of species. This information (data) can come from different sources, such as information published in print media, information catalogued in museums and information obtained directly in the field. It can also come from cartographic sources (print and digital).

2. On the one hand, biological information is required, in particular on the species or groups of fauna and vegetation as well as habitats that are considered relevant. The species to be considered for generating probabilistic models should preferably be those listed in NO M -ECO L- 059-2001, endemic species, rare species and those selected as ecologically relevant due to their role in the biological systems. The data to be collected are presented in Figure 1 indicated as biological variables. Richness and diversity data (H ') can also be collected. Habitats can be prioritised in their relevance in a categorical way to be incorporated in the models. All data should be spatially referenced (units Latitude, Longitude; U TM), using G PS.

3. On the other hand, cartographic-type information can come from digital media (such as satellite images, orthophotos, and digital charts) or images (such as thematic charts, aerial photography). It requires a software and a system that can be used to

G eographic Information Systems (Idrisi, ArcView, Ilwis, ArcInfo) to obtain the physical environmental variables of each sampling point or collection of the biological information of the relevant species or groups of species, directly from the digital and printed cartography.

4. The data of the physical and biological variables for species or groups of species are incorporated in the form of matrices where each record corresponds to a point of presence in the matrices. In these matrices the points with the absences of the species are indicated at the same time. Each matrix must be related to its corresponding spatially referenced data for each point. Probabilistic statistical models are then used, adjusting all variables to species and selecting only those that are statistically significant and have an ecological explanation according to species requirements. These final models allow predicting the distribution and abundance patterns of the species of interest (see example 1). With these models and the use of a GIS (e.g. Idrisi), species distribution maps are created.

These models are subsequently verified and validated in the field with a sampling design according to the environmental variables relevant to the species and which appear in the models. Once these data have been validated using the results of the first models, a further modelling is carried out, producing the final thematic maps.

5. With these data from the thematic maps, the areas with the highest and lowest biological diversity are identified . Subsequently, the optimisation so that, according to the restrictions and variables, the best decision is taken for the identification of priority or critical areas for the conservation of the environment.

1.9. Ecological Management

Ecological management has been defined on the basis of the experience of other countries, and has been shaped by the integration of concepts coming from d iverse scientific branches. F und am entally it has two theoretical and m ethodological sources:

1. Territorial planning. It is nourished by the conceptions of engineering and architectural planning where the key point is the "plan", the spatial location of activities. In the beginning, the so-called construction projects took into account the favourable conditions for the construction, for the settlement of the building, the industry, the housing unit; the choice of the site was made according to the favourable conditions for what was to be done, without taking into account what already existed.

   This trend has been transformed by considering it necessary to establish whether what is going to be done is appropriate to the work, to the project; it has to be considered whether the new thing, what is to be done, does not cause an impact on what already exists, both what has already been built and used by man.

2. Economic O rdinam ent. With the establishment of the need to evaluate investment costs for the location of projects and works, and with the appearance of a series of regulations for the protection of both the natural elements existing in an area and the safeguarding of acquired rights for the use and exploitation of the sites in which they are located, the need arose to give this form of planning new activities.

   What for spatial planning is the 'plan', in the case of the economic conception of 'planning', becomes a 'programme' in which the location is only one of the variables, and socio-economic, political, legal, anthropological and, in recent years, ecological considerations are added to it.

   With this new vision, the form of exploitation and use of the territory is "conditioned" to its economic and ecological viability and to its social acceptance in congruence with political tendencies and with compliance with the regulations related to it. Thus, we can say that...ecological planning is the politically and socially concerted determination of technical or technological norms of regional validity, within the framework of a land use model, which regulate and promote productive activities in accordance with the structure of the ecosystems and with the interest of current and future majorities. The following is the way in which some authors define ecological management and how they all combine the two ways of conceiving it, both at the territorial level and at the economic and environmental level. Bistrain (2000) considers that "the expression ecological management has a number of similarities with those of environmental management and spatial planning. However, they are not synonymous, because ecological management thus defined is something less than environmental management and something more than spatial planning.

Approach of the producers of the Asentam iento Cam pesino "N ueva Venezuela" for the actions of the O rdenam iento Ecológico.

Source: The author.

1.10. Types of O rders

<u>Environmental O rdenam ent</u>:

Environmental management involves the idea of planning the whole environment and its management. Perez (1999) considers in his research on environmental variable, that environmental management consists of "A concerted series of analyses,

processes and manoeuvres that allow for an adequate use of the environment, in order to promote sustainable economic development that meets the real needs of the present and future population". According to these authors, environmental management must be intimately linked to development planning and implementation by providing important inputs to carry it out.

<u>Territorial O r d e n a m m e n t</u>:

Bistrain (2000), considers that "Spatial planning refers to the definition of land uses according to its aptitudes". Spatial planning, or spatial planning, is a generic concept that refers to any official planning of a territory or area, whether it is an urban area or a regional, regional or national area.

20

In 1978, the Venezuelan Ministry of Environment and Renewable Natural Resources defined land use planning as: "the uses of the different zones that make up the national physical space, according to their intrinsic characteristics and the development objectives to be achieved within a predetermined time horizon". According to the same source, "In developing countries, where there are still large areas with low densities of occupation, spatial planning becomes one of the main environmental policies".

1.11. Objectives for the Implementation of Ecological Management

1) To have a preliminary legal instrument that indicates the attributions of the authorities in this area, as well as the possibilities of coordination, inter-federal, inter-state and inter-municipal.

2) Analyse the legal possibilities of regionalisation through decentralisation schemes by subject and function. In order to achieve this, it is necessary to

review of legislation, influencing specific project proposals in the four major regions at federal, state and municipal levels.

3) Systematise the activities to be carried out in the regions and provide the general regulatory framework for each of them in order to legally support their development and sustainability possibilities.

4) Operationalise the ecological land-use planning.

1.12. Priority Biodiversity Conservation Areas for Ecological O rdinance

The conservation of biodiversity is one of the indispensable aspects to be considered in the Ecological Management of the territory due to the potential conflicts that may arise between sectors, the one that represents the conservation of natural resources being one of the most   important for the ecological management of the country.

On the other hand, one of the principles of ecological management is to minimise conflict and maximise consensus, so the identification of priority areas for biodiversity conservation will allow in the first instance to determine the sectors with which conservation might be in conflict.

Therefore, by identifying priority areas, the best decisions can be taken on the basis of the activities carried out by other sectors and allowing them to be carried out in the most appropriate places without affecting the functionality of the ecosystems and without reducing the biological diversity of a particular area as a whole, while at the same time allowing the conservation of areas that are important in terms of the degree of diversity they contain.

In order to properly conserve biological diversity, it is necessary to locate the richest areas, those containing the greatest diversity, those with the highest biodiversity, and those with the lowest biodiversity.

The identification of critical areas for biodiversity conservation is urgent given the high rates of deforestation and changes in land use that are currently occurring, leading to the loss of biodiversity and the loss of species that are listed as having special protection status. The identification of critical areas for biodiversity conservation is urgent given the high rates of deforestation and land use changes that are currently occurring, leading to the loss of significant numbers of species, as well as of particular habitats and loss of ecosystem functionality.

Biodiversity may also be lost in a significant way, partly due to a lack of strategies to identify what is a real priority for conservation; there is an urgent need to create strategies and programmes for the conservation of natural resources. These strategies must be prioritised and be based on the systematisation of data collection and the application of a method or methods to standardise the analysis of information. The development of these strategies is being considered in the framework of the Territorial Ecological Ordering Regulation, given the pressing need to incorporate environmental sector priorities in the framework of ecological ordering.

Ecological guidelines and strategies must then consider the identification of priority areas for the conservation of biological diversity, as these areas represent potential sources of conflicts that would be generated with other land use stakeholders during the ecological management process. Within the context of the Territorial Ecological Ordinance, it would allow for the location of areas that are legally required to be conserved.

1.13. Intuitive and Quantitative Methods for Identifying Priority Areas for Conservation

Methods for identifying priority or critical areas for conservation can have different approaches, from the intuitive to the quantitative analytical; both approaches have been used for the identification of areas containing certain attributes of conservation interest.

conservation, such as the presence of flagship species, or species at risk of extinction, or the existence of particular habitats, as relevant as an oasis or gully bottoms.

Intuitive approaches, which are the ones most commonly used in analyses for the identification of critical areas for conservation, have been based on relatively simple analyses, largely based on the experience of experts. These approaches, although interesting, have a high uncertainty and a high margin of error in the delimitation of priority areas, which makes them unattractive in the context of ecological management.

On the other hand, the great development of quantitative techniques for the delim itation of critical areas for the conservation of biological diversity now makes it possible to reduce the uncertainty and inconsistency of the results. There are now different statistical approaches that allow for quantitative analyses of the biological information available for the determination of priority or critical areas for conservation. Likewise, the development of technologies for the spatial analysis of information, together with the development of statistics for spatial analysis, has allowed the generation of predictive models for the distribution of species or particular habitats at different scales.

The development of geographic information systems and statistical analyses through the generation of probabilistic models allow the analysis of information at regional scales, based on sampling methodologies at local scales. This approach is definitely the most convenient in the context of the ecological management of a region.

Predictive models can help prioritise sampling, identify priority or critical areas for conservation and improve the

conservation programmes in a region. Once the areas containing the highest specific richness of the groups of interest or where the species of conservation interest coincide (such as endemic species or those included in the protection lists of species considered key from an ecological point of view) have been located, ecological and statistical analyses can be carried out to arrive at a better design of the critical areas for conservation (Toledo, 1998).

1.14. Elem ents Relevant to Identify P rio rity Sites

It is necessary to make some considerations about the elements used to determine whether a site is a priority for conservation, for the identification of priority sites, it is necessary to define the key elements that currently should be used to consider them as a priority. Firstly, as representatives of biological diversity, all biological groups in an area should be considered and thus a quantitative, numerical measure representing all of them, but in practice this is hardly possible.

It is important to consider that species or groups of species for which there is an adequate level of information on their ecology and biology that can be relatively easily sampled should be used for the analyses and statistical modelling; this information will allow the models to be robust, and the environmental variables that appear as significant in the models can be considered as ecologically relevant variables for the species.

Secondly, it can be noted that there are two approaches to identifying priority sites for conservation; one is to consider the total richness of each biological group for which information is available and the total richness of all groups. In this way we could include the sites containing the highest biological diversity.

The other approach is to consider individual species, where only species of particular conservation interest (such as species on official protection lists, endemic species, relevant rare species, keystone species, i.e. those that are considered relevant for some of their attributes in the biological system or in the evolutionary history of the site and that are recognised nationally and internationally as species of conservation concern) can be included. In any case, be it group richness, total species richness or individual approximation of relevant species, the next step is to measure the environmental characteristics of the sites with the highest richness or relevant species in order to determine, through modelling, the environmental variables that condition the distribution of these particular groups or species. This is achieved by generating probabilistic models that essentially show the habitat selection or characteristics that are favourable to the species for their presence (PROVITA, 2009, SERM ANAT,

2010).

CHAPTER 2
M aterials and M ethods

2.1. Characterisation of the Study Area

The M unicipality of Cabim as is one of the twenty-one m unicipalities located in the state of Zulia, Venezuela. Located on the O riental Lake Coast (CO L) of M aracaibo, south of the m unicipalities of Santa Rita and Miranda, west of the state of Falcón and north of the m unicipalities of Sim ón Bolívar and Lagunillas, its capital is the city of Cabim as. The municipality is located at an altitude of 3.5 m asl and has an approximate surface area of 604 km^2 . Internally, it is divided into the following parishes: Germán Ríos Linares, Ambrosio, Carm en Herrera, La Rosa, San Benito, Arístides Calvani, Jorge Hernández, Róm ulo Betancourt and Punta G orda.

Astronomical Location: 10°28' Lat. N 70°52' Long. W -10°19' Lat. N 71°27' Long.

W - Capital: Cabim as - Surface area: 604 km - Population: 268 006 inhabitants -

Density: 443.71 inhab/km

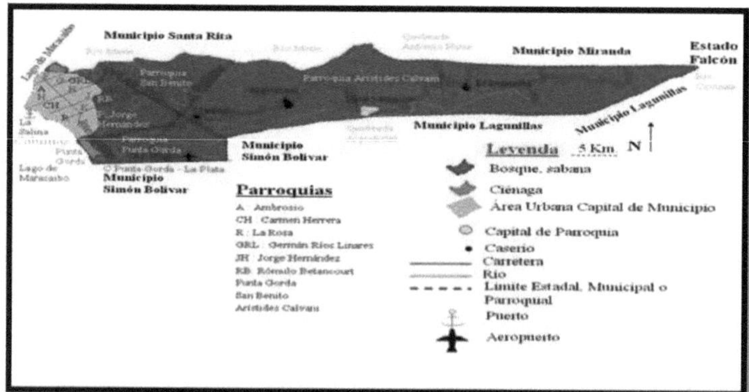

Figure 1. Municipality of Cabimas. Source: Andrés González, Wikimedia commons.

The relief is mainly flat towards the coast of Lake M aracaibo, rising towards the Lara-Zulia road, the highest elevation being the Santa Lucia hill on the border with Falcon state, with a little over 1,300 m in height. On the Lara-Zulia road, there are outcrops of the Eocene Eocene El M isoa Formation, for example in the curve of the F road and in the so-called La Antena hill.

The city has no relief and the soils have a high phreatic level, they are sands of Recent age. The Eocene outcrops are the remains of a Paleoinland, outcrops of Miocene and Pliocene age are found in the parish of Arístides Calvani. The relief in the parish of Arístides Calvani includes numerous hills and at its eastern end the Serranía de Zirum a with real mountains such as the Piñas hill of more than 1,000 m asl (Alcaldía Bolivariana de Cabimas, 2010).

2.2.1. Characteristics of the N ueva Venezuela Settlement

The farming settlement "Nueva Venezuela" is a rural farming community located on the periurban border of the city of Cabimás, in the Cabimás municipality, in the sub-region of the western shore of Lake M aracaibo. This farming settlement consists of several plots or estates with small producers of different agricultural products.

The prevailing climatic conditions in the region is a semi-arid climate, characterised by high temperatures, with an annual average of 28.3°C, and low rainfall (728.3 m m m annual average), as well as high evaporation and relative humidity (2 227 m m and 75 % annual, respectively), giving rise to very dry tropical forest and dry forest vegetation. As for the winds, the trade winds are predominant, blowing in a N E-SW direction, in the period between the months of November/April, blowing throughout the day with some lulls and a constant speed.

There are also local winds, characterised by inconsistent speeds, which occur from May to October. The origin of these winds is the result of the uneven heating and cooling of the air masses over the land surface and over the lake. Rainfall is irregular on an annual basis and there are heavy rainfalls of a hurricane-like nature (CORPOZULIA, 2011).

Figure 2. Satellite image of the rural settlement "Nueva Venezuela". Source: Google Earth.

The unit comprises a geographical area comprised of the following boundaries:

- **North:** Road G
- **South:** Highway 53
- **East:** Road F
- **West:** Sector H7
- **Parish:** Germán Ríos Linares

The unit is comprised of a population of 603 inhabitants (Pascal and Vásquez, 2012).

2.2.2. Research Variables

Biodiversity

An inventory was carried out to determine the biological diversity in the Nueva Venezuela farming settlement (M argalef, 1974).

Biodiversity or biological diversity is the variety of life. This recent concept includes several levels of biological organisation.

Com position is the identity and variety of the elements (includes what species are present and how many are present), structure is the physical organisation or pattern of the system (includes relative abundance of species, relative abundance of ecosystems, degree of connectivity, etc.) and function is the ecological and evolutionary processes (includes predation, competition, parasitism, dispersal, pollination, symbiosis, nutrient cycling, natural disturbances).

<u>Insect Diversity of the Lagoon</u>

The diversity of insects present in the community's lagoon was also monitored in order to establish, through the diversity of insects present, the type of bio-indicator that this group of fauna is, and also to deduce the conditions of the lagoon for which sampling was carried out for six months (February to July 2013).

Lagoon present in the rural community. Source: Author.

Ecological O rdenam ent

In order to implement strategies for the ecological management of the community, surveys were carried out. With this, the proposal for ecological management was made. This was done using the Participatory Action Research method. On the other hand, it can be deduced that planning is a physical, sectorial planning technique based on the incorporation of environmental and ecological variables into the process of managing human activities. It can be said, then, that it is a way of making integral a vision that for years has tried to give congruence to state and private actions around the environmental elements of the ecosystem.

2.3. Statistical Design

We worked with a sample of 10% of the total population, the statistical design was completely random, using the following statistics: mean, relative frequency (Fr) and absolute frequency (Fa). With biodiversity, an inventory of species observed for each taxonomic group of ecological importance was made, and Fr and Fa were applied to these values.

CHAPTER 3

Results and Discussion

Biodiversity is the essential and fundamental value of a natural historical process of great antiquity. For this reason alone, biological diversity has the inalienable right to continue to exist. Man and his culture, as a product, part of this diversity, must take care to protect or respect it. Moreover, biodiversity is a guarantor of well-being or balance in the biosphere. The diverse elements that make up biodiversity form true functional units, which provide and ensure many of the basic "services" for survival.

Consequently, from the human condition, diversity also represents a natural capital; in addition to use, the benefit of biodiversity has contributed in many ways to the development of human culture, and represents a potential source for meeting future needs. Considering biological diversity from the point of view of its present or potential uses and benefits, it is possible to group the arguments into three main categories.

With reference to the above, it follows that ecological restoration is an activity with different spatial and temporal scales in which anthropogenic disturbances play an important role at whatever scale it is chosen. The loss of environmental services of ecosystems is also a concern of people in any region and therefore both regional management and the needs of local communities must be taken into account; it is therefore important that people are actively involved in the formulation of restoration projects from their inception, which can guarantee their continuity and consolidation (USAID, 2006).

According to the reasoning that was made, it is expressed that in order to realise the development of the biodiversity associated with the New

Venezuela in the municipality of Cabimás, the following activities were carried out in response to the objectives set, which are described below:

Sustainable development and law are ideal foundations for understanding the processes, changes and continuities that various societies have undergone in pursuit of a balance in social, economic and anthropocentric issues related to environmental protection. Sustainable development has taken off in recent years, and at the regional level it represents a major concern, due to the loss of wildlife and natural resources that the world is facing. Law allows us to approach general patterns of social behaviour and conduct (*in abstracto*). Sustainable development, on the other hand, offers to reverse the negative trends or impacts that the globalising process of the world economy, industrialisation and technology have caused on the environment in societies (*in concreto*).

3.1. O bservations Made in the Settlement

In the different visits that were planned and carried out to the walking community, in order to meet the established objectives, it was possible to observe the biodiversity that is part of it in a preliminary way, which allowed us to be able to map it, and at the same time helped to present it as follows: First the biodiversity of the water body (flora and fauna), then the biodiversity (flora and fauna) located in the vicinity of the lagoon, then the biodiversity found 20 m away from the lagoon, and finally all the biodiversity that could be observed in the Nueva Venezuela Farmers' Settlement.

Biodiversity Study of the Lagoon

The biodiversity of the lagoon in Asentam iento Cam pesino Nueva Venezuela is made up of flora and fauna, so first the flora and then the fauna are presented (table 1).

Table 1 Biological Diversity Associated with the Settlement Lagoon Cam pesino

Common Name	Scientific name or taxon
Aquatic Plant	*Echinodorus s p .*
Water Duck	*Jacana jacana*
Iguanas	*Iguana ig uana*
B a b i 11 a s	*Caim an crocodilus*
Duckweed	*L e m n a sp.*
Caricari Eagle	*Caracara cheriway*
Dragonflies	*O donata*
G arza Real	*Ardea alba*
Fish C ac h a m a	*Colossoma m acropom un*
Bocachico fish	*Prochilodus magdalenae*
Fish T i l a p i a	*O reoch rom is sp.*
Old fish	*Aechidens pulcher*

Source: The author

Carrying out biodiversity observations together with the inhabitants of the community. Source: Author.

G a llito de agua (*Jacana jacana*) present in the lagoon of the rural settlement. Source:

Lerm ith Torres.

M onitoring the Diversity of the Lagoon

The study on the diversity of fauna (insects) found in the lagoon of the community is presented below (table 2).

Table 2 Benthic insects present in the settlement's lagoon

Group (insect)	February (abundance in d/m)	March (abundance ind/m 2)	April (abundan c i a ind/m 2)	May (abundant in ind/m)	June (abundant ind/m)	July (abundant ind/m)
O donata	3,5	1 ,5	1 ,5	3	2,5	1 ,5
C o le o p te ra (larvae)	1	1 ,5	1	2 ,5	1	1
Diptera	- -	0,5	- -	- -	--	- -
H e m m i p t e	- -	0,5	- -	- -	- -	- -

r a						

Source: Author, 2017.

It is noted that during this monitoring the order of Odonata (dragonfly larvae) can indicate good conditions of the aquatic ecosystem. From this point of view it is feasible that the water body is in oligotrophic to mesotrophic conditions (Alba and Sanchez, 1988). In this way, the aquatic insects provided "biological services" to the inhabitants of the community, as this allowed inferences to be made about the conditions of the water body.

Biodiversity sampling near the lagoon area of the "N ueva Venezuela" settlement Source: Author.

A lagoon within the cam pesina com m unity, covered with the aquatic plant *Lem na sp.* Source: Author.

Terrestrial species of flora and fauna found in the Nueva Venezuela farming settlement.

For this purpose, the species found at a distance of 20 m from the lagoon of the Asentam iento Cam pesino N ueva Venezuela were taken into account. See tables 3 and 4 respectively.

Table 3. Plant Species in the Near Lagoon Area

Common Name	Scientific name or taxon
G u i n e a	. P a nicu m m axim u n .

Herbaceous	Viola riviniana
Black olive tree	Cappa ris odoratiss im a
Cactus	Cactaceae
San Francisco Tree	Luehea divaricata

Source: The author

Table 4. Animal Species in the Near Lagoon Area

Common Name	Scientific name or taxon
Termits	C ry to p te rm s
Orange-legged spider	Achaearanea tepidariorum
Hormigas	A tta s p.
Dragonflies	Odonata

Source: The author

It should be noted that the species described above (tables 3 and 4) are found in the surroundings of the com m unity's lagoon, in a tropical dry forest ecosystem. The Formicide observed in this monitoring (*Atta sp.*) has been involved in pest problems, according to the inhabitants of the com m unity. The leaf-cutting ant or bachaco performs a pruning action on some crops, reducing the photosynthetic capacity of the plants (Pascal *et al.*, 2015).

Table 5. Other species identified in the Near Lagoon Area.

Common name	Scientific name or taxon
Cuji	Prosopis juliflora
Cigar	Xylocopa violacea
Mariposa	Lepidoptera
Snail	M olusca - Pom acea
Canines	C anis fam ilia ris
Bees	Apis melifera
Dung beetle	C o le o p te ra
G usano verde	Hypera postica
Turpial	Icterus ic te ru s

Source: Author.

Wild Vertebrates Present in the Settlement

Wild mammals are listed in table 6.

Table 6 Wild mammals within the Campesino Settlement

Common name	Scientific name or taxon
Rabipelado	D id e lp h is m arsup ia lis
Cachicam or	D asypus novencitus
Mapurite	C oenepatus s e m is tria tu s
Rabbit	Ssyvilagus floridanus
Anteater	M ym ecop hag a tr id a tr id a c ty la

Source: The author

Amphibians are considered to be animals of great ecological importance due to their role in ecosystem food chains and their contribution to regional biodiversity. On the other hand, some species behave as crop pests, as well as being potential reservoirs and transmitters of diseases to humans. It is worth noting that studies of mammalian units in the country are scarce (Belandria, 2009).

Birds of Ecological Importance

Birds of ecological importance that were sighted at the Asentam iento Cam pesino Nueva Venezuela were a very diverse

group (table 7).

Table 7 Birds of Ecological Importance within the Nueva Venezuela Settlement.

Common name	Scientific name or taxon
Z a m uro (Vulture)	*C orag yp s a tra tu s*
Alcaravan	*Vanellus chilensis*
P a lo m a sabanera	*Zenaida a u ric u la ta*
Royal parrot	*Am azona ochrocephala*
Dirty-faced parakeet	*A ra tin g a pe rtin a x*
Cristofué	*Pitangus sulphuratus*
Paraulata llanera	*M im us g ilv u s*
Tile	*Thraupis episcopus*
T u r p i a l	*Icterus ic te ru s*
Chocorocoy	*Campylorhynches nuchalis*
Atrapam oseas bull's blood	*Pyrocephalus rubinus*
Common cockroach	*Troglodytes aedon*
Caricari Eagle	*C a ra c a ra che riw a y*
Sparrowhawk prime ito	*Falco sparverius*
Carpenter	*Chysoptiilus zuliae*
Palom ita m araquera	*Columbina squamata*
G o n z a l i t o	*Icterus n ig ro g u la ris*
Hummingbird	*Lkeucippas cemina*

Source: Author.

Turpial (*Icterus icterus*). Source: Lermith Torres.

Birds are of considerable ecological importance in the community, as they provide a large number of biological services,

and some of them even act as biological controllers in these agroecosystems by feeding on insects. Similarly, birds can

indicate determining conditions in the environment, as some are susceptible to drastic changes in the ecosystems.

Bull's blood traps (*Pyrocephalus rubinus*). Source: Lerm ith Torres.

Many bird species have been affected by increasing deforestation to create space for conventional livestock and agriculture, which is why it is necessary to carry out these types of inventories in agricultural communities (Hernández, 2009).

Garden tile (*Thraupis episcopus*). Source: Lerm ith Torres.

Reptiles and Amphibians

Among the Herpetofauna found in Asentam iento Cam pesino Nueva Venezuela, there is a great variety of species as shown in table 8.

Table 8 Reptiles and Amphibians present in Asentam iento Cam pesino Nueva Venezuela

Common name	Scientific name or taxon
Rattlesnake	*C ro ta lu s d u ris sus*
Coral	*Micrurus dissoleucus*
T ragavenado	*Boa cons tric to r*
Buzzard snake	*Mastigochyas pleei*
Common bush	*Am eiva am eiva am eiva*
B a b illa	*Caim an crocodilus*

Morrocoy turtle	*Geochelone carbonaria*
Tuqueque (G eko)	*Phyllodactylus dixoni*
Iguana	*Iguana iguana*
Lizard	*Anolis sp.*
Icotea turtle	*Geochelone carbonaria*
Common toad (amphibian)	*Bufo marinus*

Source: Author.

The Herpetofauna present in the Nueva Venezuela farming settlement is of great ecological importance, however,

hunting and intervention in the nests of these vertebrates can produce a drastic decrease in their population levels,

which could have negative repercussions on the ecosystem (PROVITA, 2009).

B abilla (*Caim an crocodilus*) in the lagoon, only the cephalic region is visible, in the central area of the
image. Source: Author.

Plant and Forest Species of the Cam pesino Nueva Venezuela Settlement

Tall trees with large foliage were observed in the com m unity, which at first glance appeared to be free of phytosanitary

problems, indicating that the environmental conditions were suitable for each species (table 9).

Table 9 Forest plant species present in the farming settlement

Common name	Scientific name or taxon
Naked Indian	*B u rs e ra sim aruba*
J a b i 11 o	*Hura crepitans*
Black olive tree	*C apparis o d o ra ra tis s im a*
O livo sagrado	*Capparis liearis*
Peresquia	*Leafscale gull Rhodocaptus*
Dividive	*Caesalpinia coriaria*
Vera	*B ulnes ia arborea*
Cardon	*Stenocereus griseus*
Jebe	*Lonchocarpus a tro p u rp u re u s*
Jobo	*Spondias mombis*
C a u j a r o	*C o rd ia denta ta*
Black cují	*Acacia macracanta*

Source: Author.

M onitoring of the vegetation present in the com m unity cam pesina. Source: Author.

The presence of trees is fundamental in any ecosystem (M argalef, 1974), including agroecosystems, as they offer a great diversity of biological services such as natural barriers, proportion of ecological niches, among others. In the community they could be used as a basis for the use of agroforestry (Altieri, 2000).

Main vertebrate groups by number of species reported in the Asentam iento Cam pesino Nueva Venezuela

The groups of vertebrates identified in the Nueva Venezuela farm settlement show a greater number of species identified in the class: Birds (18 species identified).

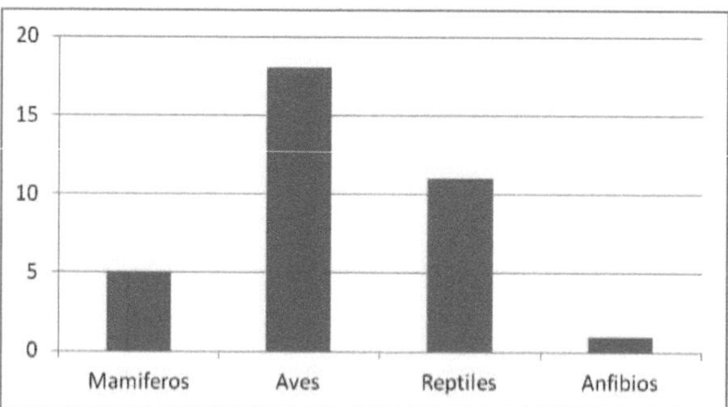

Figure 3. Main vertebrate groups reported by number of species, in the farming settlement.

Relative and absolute frequencies of the fauna identified in the unit

When the inhabitants were consulted on the sighting of fauna, it was possible to prove what was noticeable in each of the visits, such as the fact that the highest percentage was that of birds with a value of 51.43%, followed by reptiles with 31.93%, then amphibians with 14.28% and lastly amphibians with a value of 2.85% (table 10).

Table 10 Relative and absolute frequencies of the fauna identified in the com unit

Biodiversity	F recency	Frequency

	Relative (Fr)	Absolute (Fa)
M a m e m b e r s	5	14,28
Aves	18	51,43
Reptiles	11	31,93
Amphibians	1	285
Total	35	100

Source: The author

Birds are of ecological and cultural value to humans, they are good indicators of the state of conservation of the environment, through their study, we can better understand the changes that are affecting the environment, providing a useful means to improve scientific knowledge and understanding of environmental issues.

In relation to the topics discussed, the aim is to reflect on the importance of protecting birds and their habitats for the sake of the general health of the environment, to develop ideas and possible solutions for the purpose of conserving this group of animals, and to develop ideas and possible solutions for the purpose of conserving this group of animals. Any disturbance of the delicate balance of ecosystems can result in the disappearance of numerous species that depend on them, as well as having serious consequences for the human race. The conservation problems faced by birds within the ecosystem under study are very important: habitat degradation, urban sprawl, indiscriminate hunting and many others related to human population growth. Even solid waste left by people in their activities can attract predators of these birds. Although these vertebrates showed the greatest number of species, they are very vulnerable animals due to their gregarious behaviour (Piña *et al.*, 2014).

Ecological Development Proposal for the Nueva Venezuela Rural Settlement

The development of strategies or programmes for land settlement should be considered under the Ecological Land Use Planning theme, given the urgent need to incorporate environmental sector priorities into the ecological planning framework. Ecological guidelines and strategies must then consider the identification of priority areas for the conservation of biological diversity, as these areas represent potential sources of biodiversity.

The conservation of agricultural ecosystems or agro-ecosystems is of great importance in the context of land-use planning. The conservation of agricultural ecosystems or agro-ecosystems is of great importance for the food security of the planet, as they contain the necessary elements (soil, water and biodiversity) that are consubstantial to agricultural production.

Although agriculture is a necessary form of land use that is in opposition to ecosystems, because decisions on land use and management practices influence ecological processes and soil-water-plant interactions, these decisions must take into account that people's quality of life and well-being ultimately depend on the well-being of the ecosystem (Machado and Cam pos, 2008).

Once the biodiversity of the rural community was known, the villagers were interviewed in order to make a contrast with what was observed in the field.

According to the interviews carried out with the inhabitants of the Nueva Venezuela Settlement, regarding the number of species, the citizens affirmed that the greatest number of species is found in the lagoon and on the borders of the unit, this is feasible due to the fact that the lagoon offers other ecological niches for the biota, apart from the terrestrial system, in addition to obtaining water which is very important for any biological entity. Similarly, the study community is adjacent to other rural areas, which significantly reduces human intervention and increases the possibility of finding more biological diversity in these areas.

In view of the above, it can be seen that there is the greatest diversity in the limits of the community, with a total of 50 %, in the surroundings of the lagoon 33.33 %, on the main access road 6.67 % and finally, on the farms or plots of land 10 %. See table 11

Table 11 Relative and absolute frequencies of the most biodiverse sites in the com m unity according to the villagers

Answers	F r	F a
Limits of the unit	3 0	5 0
Lagoon	2 0	3 3,33
Main access road	4	6.6 7
Farms or plots of land	6	10

Source: Author.

It should be noted that during the biodiversity inventory carried out, the species identified were mostly observed in the areas described by the community, i.e. the boundaries of the settlement and the lagoon.

Similarly, the inhabitants of the rural settlement were asked which components of the fauna they considered to be the most abundant, to which the majority responded birds with a value of 53.33%, mammals 33.33%, reptiles 8.33% and lastly amphibians with a value of 5% (Table 12), coinciding with the biodiversity inventory carried out.

Table12 Relative and Absolute Frequencies of the Most Abundant Fauna according to the Community

Answers	F r	F a
Birds	3 2	5 3,33
Mammals	2 0	3 3,33
Reptiles	5	8,3 3
Amphibians	3	5

Source: Author.

However, according to direct observations and the identification of species present in the unit, the second most abundant group in terms of number of species is that of reptiles, and the third most abundant group is that of amphibians (Table 12).

It is also important to note that most of the inhabitants of the rural settlement "Nueva Venezuela" are not aware of the importance of the biodiversity associated with the community, and there was also a lack of knowledge of what an Ecological Management Plan is and the benefits it could bring to the community (Table 13).

Table 13 Knowledge of Villagers on the Importance of Biodiversity and Ecological O rdenam ent.

	F r	F a	F r	F a
Awareness of the importance of biodiversity	S i: 1 9	Si: 3 1,66	N o : 4 1	No: 68,33
Knowledge of an Ecological O rdinance	Si : 8	Si: 13,33	No: 52	No: 86,66

Source: The author

Based on the observations of the biological diversity present in the community and what was established by the inhabitants of the community, it was estimated that there are some important strips of biota, which comprise two important zones: the limits of the community and the lagoon that is within it. Therefore, it is important for the ecological management that these zones are only for the protection of biodiversity, and therefore they were advised not to practice agricultural activities in these important zones. On this basis, it was determined that any activity should be carried out within a distance of approximately 500 m from the lagoon, thus giving it a position to be used as a buffer zone.

The perimeter of the settlement should also not be encroached upon.

Current Situation of the Ecological O rdenam ent and Identification of the Biodiversity of the Farm Settlement

Traditional agricultural activities have been developed with their back to the conservation of the environment, biodiversity and social differences. It is necessary in the productive rural sectors to carry out an ecological territorial planning that benefits the community, the existing biodiversity in the sector, and to carry out sustainable and environmentally friendly agricultural activities. The rural settlement of Nueva Venezuela has sufficient land and water resources to be able to develop agricultural and livestock plans that are environmentally friendly, but there is a lack of knowledge about biodiversity and ecological planning.

This research was aimed at highlighting the importance of recognising natural spaces in the current global context of socio-spatial and territorial changes, based on the experience of observation, the aim is to analyse the state of biodiversity in the "Nueva Venezuela" Rural Settlement and to determine how it can be used.

In addition, to be able to detect the benefits that the application of ecological agriculture would have for the community, as it will allow them to develop a tool for the supply of food produced by themselves, and also to inform the inhabitants of the great importance of the existing body of water in the locality, as it can generate benefits at an economic level in terms of the use of aquaculture.

The analysis of a territorial space is complex, based on the fact that there are different approaches to its study. For the development of this research, special attention has been paid to the settlement of populations in forested areas and other areas of globalisation, and it cannot be otherwise, without analysing the territory under study as an isolated space; rather, it should be analysed from the different scales and areas that integrate and influence it. There are areas of the world that are completely abandoned to their fate, the demographic increase in population, social, environmental or economic problems have led to the population of environmental areas where there is an impact on the characterisation of the environment and its natural components (G óm ez, 2002).

Territory is an important element in identity, because it locates populations, offers possibilities, establishes certain conditions. The interaction between society/territory over time gives rise to areas with very unique ways of life, customs and organisation of space; the city of Cabimbas is a sample of the different settlements that man has made throughout its extension in the different sectors, neighbourhoods, communities and urbanisations.

Ignorance of the biodiversity existing in a territory for agricultural use could trigger its loss and the benefits that biological diversity could have in an integrated agricultural system; however, the lack of ecological management places this biodiversity at risk, and this is a fundamental aspect to avoid potential conflicts that could arise in the community. The immediate need to make a rational use of environmental resources and the urgent need to stop the deterioration and degradation of ecosystems in Latin America, specifically in Venezuela, makes the study of ecosystems a priority in order to propose better spatial planning within a socio-economic and ecological environment (Berroteran, 2001).

According to the deductions made above, it can be deduced that the following situations were specifically detected: lack of knowledge of the biodiversity that inhabits the unit, lack or absence of ecological territorial planning, little knowledge of environmental and agro-ecological issues, poor conditions of access roads to the unit, lack of some

Basic services, such as electricity, domestic gas and drinking water, and the poor state of the houses in the community were the problems with the highest priority in the selected community that need to be solved. Through the implementation of ecological planning and the identification of biodiversity in the Nueva Venezuela farming settlement, it will be possible to make sustainable and equitable use of the territory of the com m unity. from this sustainable point of view, the complexity and stability of agricultural systems, similar to that of natural systems, is based on their diversity. This diversity, constituted by a mosaic of elements - the agricultural landscape - linked by a series of flows (materials, energy, organisms, etc.) horizontally between them and vertically within each of them in interaction with the local use of the resources of the rural culture in the community, is the basis for sustainable agrosystem management and the design of practices that maintain or increase the fertility, productivity and quality of productions and regulate pest populations (Sans, 2007).

CHAPTER 4

Conclusions

- The rural settlement Nueva Venezuela has a lagoon which represents an important reservoir of biological diversity, and also because of the diversity present in   it, it denotes conditions of macro-trophic to oligotrophic, so it could have other uses, finding species of food interest.

- The group of wild vertebrates with the greatest number of species found was that of birds.

- The areas of greatest biological diversity observed were in the The lagoon surroundings and the perim etral area of the com m unity cam pesina were therefore proposed as biota safeguard zones for the settlement's ecological management.

ANNEX : Interview for the ecological management proposal.

Where do you think you see the most wildlife species?

Lagoon: _______________ Finca: _____________________ Main road:_______________

Perimeter of the settlement: ___________________

What do you think is the most abundant animal group?

M amm ers: _____________ Birds: _______________Reptiles: ______________Amphibians: _______

Do you know the importance of Biological Diversity for the Community?

Yes: __________ No: ____________________

Do you know what an Ecological Order is?

Yes: __________ No: _______________

Bibliographical references

1. Alba-Tercedor, J. and Sánchez-O rtega, A. (1988). **A rapid and simple method for assessing water quality based on that of Hellaw ell (1978).** *Kimnetica* 4: 51-56.

2. ALCALDÍA BOLIVARIANA DE CABIMAS (2010). **Geography of the Municipality of Cabimas.** [Text

online   ]   available at:
www .alcald iabolivarianad ecabim as.gob.ve . [Accessed: 2013, July 15].

3. Altieri, M . Nicholls Clara (2000). **Agroecology: Theory and Practice for Sustainable Agriculture.** UNEP (United Nations Environment Programme). First edition. ISBN : 968-7913-04-X. M exico.

4. ADC Corp (2000). **Flora and Fauna Inventory.** Technical report for the M onte Sierra residential development project. M ayagüez, Puerto Rico.

5. Barrera A. (2012). **The Protection of Wild Species. Especial Tratam iento de la Protección *in situ*.** Doctoral thesis. University of Alicante, Spain.

6. Belandria, A. (2009). **Estudio de una Com unidad de Pequeños M am íferos Am íferos Am enazada de Extinción en una Zona Agrícola del Sur del Lago deM aracaibo.** Universidad de los Ande (ULA). P ROVITA. A Hand to Nature: Conserving   Endangered Species Venezolanas. Legal Deposit: If25220095744168. ISBN: 978-980-6774-032. Caracas, Venezuela

7. BERRO TERAN (2001). **Enfoque M etodológico de O rdenam iento Ecológico.** [Online text] available at: http://www2.ine.gob.m x/publications/books/338/Berroteran.pdf [C o n s u l t : 2012, Ju l y 10].

8. . B istra in , G u illerm o . (2000) **Instrum entos para el Desarrollo Urbano** . UNAM (National Autonomous University of Mexico). PUEC. M exico.

9. Boada, C. Cam paña, J. (2008). **Com position and Diversity of Flora and Fauna in the Carchi Province.** Report of ecological assessments. EcoCiencia. ISBN: 978-9942-01-697-3. Quito, Ecuador.

10. CAMPOS MAYBE, MACHADO HILDA, MATÍAS YANETSY, G O NZÁLEZ LEYBIZ, SÁNCHEZ SARAY AND DUQUESNE P. (2005).   **Diagnosis socio-economic, environmental and institutional aspects of a productive entity through participatory methodologies.** *Pastures and Forages*, Vol. 28, No. 4

11. CARMONA, MARÍA (2011). **Boletín Com parado de Derecho Mexicano.** Biblioteca Jurídica Virtual (Online text). Available at: http://www.juridicas.unam .m x/publica/rev/boletin/cont/78/art/art2.htm.

12. Casler, C. (1992). **Ornithological Bibliography of Venezuela.** *Boletín del Centro de Investigaciones B io ló g ic a s* . Vol. 23. N° 1. Universidad del Zulia. M aracaibo, Venezuela.

13. Corporation for the Development of the Zulian Region - CORPOZULIA, (2011). **Characteristics of the M unicipality of Cabim as.**

14. Crem one, C. Cervigón, F. Gorzula, S. M edina Glenda, Novoa, D. (1986). **Fauna of Venezuela, Vertebrates.** Editorial Biosfera. Caracas, Venezuela.

15. DUNCAN, D.B.(1955). **M ultipleRange and M ultiple F. tests**. Biometrics.11: 1

16. Delgadillo, J. and Torres, F. (2008). **M ulticausal dimensions of territorial planning:** Política Territorial en M exico. Hacia un m odelo de desarrollo basado en el territorio, Javier Delgadillo (coord.), Sedesol, Instituto de Investigaciones Económ icas UNAM , Plaza y Valdés, M exico.

17. Escobar, A. (1997). **Cultural politics and biological diversity: State, capital and social m ovem ents in the Pacific coast of Colom bia,** (M imeo.), Department of Anthropology, University of M assachusetts, Am herst.

18. Gil Karine, Casler, C. W eir, E. (2003). **Biodiversity in Lake Maracaibo.** Ediciones Astro Data, S. A. ISBN: 980-232-883-9. Maracaibo, Venezuela.

19. GROBET, G. (2003). **Annual M ethods of M ulti-objective M ulticretrial Decision Analysis Methods for Ecological M anagement**. Final Project Report. Mexico D. F.

20. G óm ez, D. (2002). **Ordenación Territorial**. Ediciones M undiPrensa. Editorial Agrícola Española, S. A.

21. G onzález, J. Pascal, E. Vásquez Helim ar (2014). **Gestores Am biental Com buntarios para la prom oción de la Cultura Am biental, Parroquia francisco O choa, M unicipio San Francisco**. *Refereed proceedings of the I International Congress for Endogenous Development*. Centro de Investigación para el Desarrollo Endógeno (CIPD E) UNE RM B. ISBN: 978980-6792-22-7.

22 Harper, J. (1977). **Population Biology of Plant**. Academ ic Press, New York

23 Hernández Laura (2009). **Ecological and Genetic Aspects of Endemic and Threatened Bird Species.** Universidad Central de Venezuela (U CV). PROVITA. Una M ano a la Naturaleza: Conservando las Especies Am enazadas Venezolanas. Legal Deposit: lf25220095744168. ISBN : 978-980-6774-03-2. Caracas, Venezuela.

24 Hernández-Sam pieri, R. Fernández, C. Baptista Pilar (2006). **M etodología de la Investigación**. Editorial McGraw-Hill. ISBN: 970-10-5753-8. M exico.

25 Lentino, M. (1997). **L ista Actualizada de las Aves de Venezuela.** Catálogo Zoológico de Venezuela. Vol. 1. Museo de Ciencia y Tecnología de M érida, Venezuela.

26 .Llam ozas, R. D uno de Stefano,W . M eier, R. Riina, F. Stauffer, G . Aym ard, O . Huber and R. O rtiz (2003). **Libro Rojo de la Flora Venezolana**. Caracas, Venezuela: Provita, Fundación Polar, Fundación Instituto Botánico de Venezuela Dr. Tobías Lasser. 555 p.

27 M assiris, A. (1993). **Bases Teórico-m etodológicas para estudios de O rdenam iento Territorial**. IDCAP, M ission Local, Instituto de Desarrollo del Distrito Capital y la Participación C iudadana y Com unitaria ID CAP, Year 2, No. 2, January/March, Universidad Distrital, Santa Fe de Bogotá.

28 .MACHADO HILDA, SUSET A, MIRANDA TAYMER, CRUZ AIDA, OLIVERA YUSEIKA, MILERA MILAGROS, CAMPOS MAYBE AND DUQUESNE P. (2007). **Local development management in municipalities: the municipal initiative as an experience of change in the province of M atanzas**. *Pastures and Forages*, Vol. 30, Special issue, p. 45.

29 .M ARGALEF R. (1974). **Ecología**. Ediciones O m ega, S. A. Barcelona, Spain.

30 . M A R N - Ministry of Environment and Renewable Natural Resources (2003). **M em ory and Account**, Caracas, pp. 60-65 (Online text). Available at: www.marn.org.ve/m a r n /.

31 M oreno Claudia (2001). **M ethods for M easuring Biodiversity**. ORCYT-UNESCO. ISBN: 84 - 922495 - 2 - 8. Zaragoza, Spain.

32 Narvaez Eliana, Bautista Cecilia (2012). **Preliminary Inventory of Fauna and Bryoflora Associated to Finca los Alpes.** Faculty of Exact, Physical and Natural Sciences. University of Santander (UDES). Colom bia.

33 . PASCAL E, VÁSQUEZ HELIMAR, ALASTRE MARCELINA (2015). **Educational Activities for the Study of Biodiversity in Different Sectors of the C.O.L.** *Escenario Educativo* . Vol. 1, N° 1. January-June: 114 - 126. Legal Dep.: pp 201502ZU4604. ISSN: 2443-4493. Centre for Educational Research (CIE) UNERMB.

34 .Pascal Edison (2017). **Aquatic insects as indicators of water quality in a lagoon with aquaculture potential**. Editorial Académ ica Española (EAE) ISBN: 978-3-659-65850-1 . Saarbrücken, Germany. www .eae-publishing.com .

35 Pascal, E. Vásquez Helim ar, Pozo, J. Tim aure, C. Da Costa Nayeska (2015). **Propuesta para el M anejo Agroecológico de *A tta sp* en una Com unidad Cam pesina del M unicipio Cabim as, Zulia.** Proceedings of the XXIV Venezuelan Congress of Entom ology. Area: Control and integrated pest management. Universidad Centro Occidental "Lisandro Alvarado" UCLA. Barquisim eto, Venezuela.

36 .PASCAL E, VASQUEZ HELIMAR (2012). **Unerm b responds to the Com m unities: Estudio Am biental de un Cuerpo de Agua ubicado en el Asentam iento Cam pesino Nueva Venezuela**. UN ERM B News. Year 1 , N° 13, pg. 9. C abim as, Zulia State.

37 Pascal, E. Villarreal, A. Vásquez Helim ar (2016). **Flora and Fauna Inventory in the V itrin a Area, ZICO LCA**. Research Project. Centro de Investigaciones Educativas de Biología y Q uím ica (CIEBYQ). UNERMB.

38 . Pascal, E. Vásquez, H. Chirinos, A. San Blas, E. (2016). **Agroecology and Pest Insect Management.** *Memorias Arbitradas de las IV Jornadas Científicas del Departam ento de Ciencias Naturales*. Paper. ISBN: 978980-6792-68-5. Universidad Nacional Exp. Rafael M aría Baralt (UN ERM B). Cabim as, Venezuela.

39 Pérez, A. (1999). **La Variable Am biental Urbana: Nociones Generales y Ám bitos de Aplicación en Venezuela.** *Revista Geográfica Venezolana.* V o l. 40. N° 2. 201 - 210.

40 Piña, R. Torres, L. Corso, L. and Pascal, E. (2014). **O rnitofauna Presente en el Ecosistem a de M anglar del Parque La Laguna Azul, M unicipio Cabim as.** *Mem ories of the V Venezuelan Congress on Biological Diversity*, "Lands and Territories for the Defence of Life". Pg. 32. Universidad Bolivariana de Venezuela (UBV), M aracaibo, Zulia.

41 .PROVITA (2009). **U na M ano a la Naturaleza: Conservando las Especies Am enazadas Venezolanas.** Legal Deposit: If25220095744168. ISBN: 978-980-6774-03-2. Caracas, Venezuela.

42 . PLONCZAK MIGUEL. (1999). **El O rdenam iento Territorial y la Conservación de la Biodiversidad en Venezuela: Una Propuesta para los Llanos Occidentales.** *Revista Geográfica Venezolana.* Vol. 40. N° 1. 65 - 73.

43 Polanco, R. (2002). **El M anejo de la Fauna Silvestre, su Sostenibilidad y los Congresos Realizados en el Tema a.** Sustainable use within an ecosystem approach. IUCN.

44 Ram o Cristina, Ayarzaguena, J. (1983). **Fauna Llanera. Apuntes sobre su M orfología y Ecología.** Cuadernos Lagoven. Lagoven Public Relations Department.

45 . RO DRIG UEZ, R. (2010). **M anual de M étodos para Identificar Áreas de Conservación de la Biodiversidad para el O rdenam iento Ecológico.** M anual realizado para el SE RM ANAT. M exico D. F.

46 Rodríguez, J.P. & F. R ojas-Suárez (1998). **Endangered fauna of Venezuela: past causes, current pressures and future perspectives.** *Vida Silvestre Neotropical.* Vol. 7(2-3): 90-98.

47 Rom ero, A. (2003). **D eath and taxes: the case of the depletion of pearl oyster beds in the sixteenth-century Venezuela.** *Conservation Biology.* Vol. 17: 1013-1023.

48 Sans, F. (2007). **The Diversity of Agroecosystems.** *Ecosistemas.* Vol. 16. N° 1. Spanish Association of Terrestrial Ecology (AEET). www.revistaecosistemas.net.

49 . SARPA-Servicio Autónom o de los Recursos Pesqueros y Acuícolas (1996). **Estadísticas del Subsector Pesquero y Acuícola de Venezuela.** Inform e Técnico. M inistry of Agriculture and Breeding. Caracas, Venezuela.

50 Sarandón, S. (2009). **Biodiversity, Agrobiodiversity and Sustainable Agriculture.** Latin American Scientific Society of Agroecology.
Strands of Agroecological Thinking: Foundations and Applications. Editor/Com p ila t e r: M iguel Altieri. M e d e llín , C o lo m bia. www .agroeco.org/socla

51 .SECRETARIAT OF ENVIRONMENT AND NATURAL RESOURCES. SERMANAT (2009). **G uía de ordenam iento ecológico del territorio para autoridades m unicipales.** D irección G eneral de Política Am biental e Integración Regional y Sectorial. M exico D. F.

52 .SECRETARÍA DE MEDIO AMBIENTE Y RECURSOS NATURALES, M EXICO . SERM ANAT (2010). **O rdenam iento Ecológico**. [Text online] available at:

http://www.sem a r n a t. g o b .m x/te m m as/ecologicalp ayments/Pages/O r d e n a m m e n t E co l% C 3% B 3 gico .aspx . [Accessed: 2012, July 10].

53 Soriano, P. O choa, J. (1997). **Lista Actualizada de los M am íferos de Venezuela.** Zoological Catalogue of Venezuela. Vol. 1 . M useo de Ciencias y Tecnología de M érida, Venezuela.

54 Snyder, F. Steven, R. Beissinger, N. (1987). **New W orld Parrots in Crisis: Solutions for Conservation Biology.** Research Gate. Available at   :

https://www.researchgate.net/researcher/56288616 Noel F R Snyder

55 Toledo, A. (1998). **Economics of Biodiversity.** UNEP (United Nations Environment Programme). ISBN 968-7913-02-9. M exico.

56 IUCN-International Union for Conservation of Nature (2009). **Red List of Threatened Species. Biodiversity at Risk.** (Online text. Available at: https://elblogverde.com /lista-roja-uicn-2009-de- especies-am enazadas-biodiversidad-en-peligro/

57 USAID-United States Agency for International Developm ent / Honduras (2006). **Integrated Management of Environmental Resources.** International Resources Group. W ashington, D.C.

58 .ZENTELLA, J. BAUTISTA JESSIKA, MORALES JOSEFINA. (2010). **Guía M etodológica para Elaborar Program as M unicipales de O rdenam iento Territorial.** Secretaría de Desarrollo Social. M exico D. F.

59. Vásquez, H. (2011). **La Educación Am biental com o Eje Transversal en la Form ación de Profesionales.** Presentation at the I Cycle of Conferences "Form ando Triunfadores Integrales" M isión Sucre, Universidad Bolivariana de Venezuela. M aracaibo.

yes
I want morebooks!

Buy your books fast and straightforward online - at one of world's fastest growing online book stores! Environmentally sound due to Print-on-Demand technologies.

Buy your books online at
www.morebooks.shop

Kaufen Sie Ihre Bücher schnell und unkompliziert online – auf einer der am schnellsten wachsenden Buchhandelsplattformen weltweit! Dank Print-On-Demand umwelt- und ressourcenschonend produzi ert.

Bücher schneller online kaufen
www.morebooks.shop

Printed by Books on Demand GmbH, Norderstedt / Germany